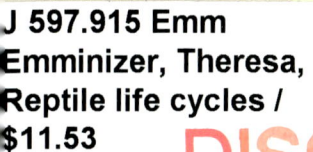

REPTILE
LIFE CYCLES

BY THERESA EMMINIZER

Enslow
PUBLISHING

Please visit our website, www.enslow.com. For a free color catalog of all our high-quality books, call toll free 1-800-398-2504 or fax 1-877-980-4454.

Cataloging-in-Publication Data
Names: Emminizer, Theresa.
Title: Reptile life cycles / Theresa Emminizer.
Description: Buffalo, NY : Enslow Publishing, 2025. | Series: Life cycles in review | Includes glossary and index.
Identifiers: ISBN 9781978539853 (pbk.) | ISBN 9781978539860 (library bound) | ISBN 9781978539877 (ebook)
Subjects: LCSH: Reptiles–Life cycles–Juvenile literature.
Classification: LCC QL644.2 E535 2025 | DDC 597.91'56–dc23

Published in 2025 by
Enslow Publishing
2544 Clinton Street
Buffalo, NY 14224

Copyright © 2025 Enslow Publishing

Portions of this work were originally authored by Bray Jacobson and published as *A Look at Life Cycles: Reptile Life Cycles*. All new material in this edition is authored by Theresa Emminizer.

Designer: Leslie Taylor
Editor: Theresa Emminizer

Photo credits: Cover (photo) DariaGa/Shutterstock.com; series art (blue/green background) Denis Ka/Shutterstock.com; series art (grunge shapes) Midstream/Shutterstock.com; p. 5 Elana Yakubov/Shutterstock.com; p. 7 (turtle) Michal Dobes com/Shutterstock.com, (tuatara) BeautifulBlossoms/Shutterstock.com, (snake) Worraket/Shutterstock.com, (crocodile) worananphoto/Shutterstock.com; p. 9 (top and bottom) udaix/Shutterstock.com; p. 11 David Evison/Shutterstock.com; p. 13 COULANGES/Shutterstock.com; p. 15 Catchlight Lens/Shutterstock.com; p. 17 Mark_Kostich/Shutterstock.com; p. 19 Xben911/https://commons.wikimedia.org/w/index.php?title=File:Jonathan-plantation-house.jpg&oldid=789232932; p. 21 (top) Natalia Kuzmina/Shutterstock.com, (bottom) Rudmer Zwerver/Shutterstock.com; p. 23 slowmotiongli/Shutterstock.com; p. 25 Rich Carey/Shutterstock.com; p. 27 White Space Illustrations/Shutterstock.com; p. 29 Joe Farah/Shutterstock.com; p. 30 Vector Tradition/Shutterstock.com.

All rights reserved. No part of this book may be reproduced in any form without permission in writing from the publisher, except by a reviewer.

Printed in the United States of America

CPSIA compliance information: Batch #CSENS25: For further information, contact low Publishing at 1-800-398-2

Find us on

Contents

Meet the Reptile! 4
All Kinds of Reptiles 6
A Reptile's Life 8
Laying Eggs 10
Incubation 12
Hatchlings 14
Juveniles 16
Adulthood 18
Live Birth 20
Life of a Green Anaconda 22
Aquatic and Semi-Aquatic 24
A Sea Turtle's Life 26
Staying Warm 28
The Basic Reptile Life Cycle 30
Glossary 31
For More Information 32
Index 32

Words in the glossary appear in **bold** the first time they are used in the text.

Meet the Reptile!

A reptile is a cold-blooded animal. That means their body **temperature** changes based on the temperature around them. Reptiles breathe air. They are also vertebrates, which means they have backbones. Their skin is covered with hard plates called scales.

LEARN MORE

REPTILES HAVE LIVED ON EARTH MUCH LONGER THAN PEOPLE HAVE. SOME OF THE EARLIEST-KNOWN REPTILES LIVED OVER 300 MILLION YEARS AGO! DINOSAURS WERE REPTILES THAT WENT EXTINCT, OR DIED OUT COMPLETELY.

Komodo dragon

All Kinds of Reptiles

There are thousands of different species, or kinds, of reptiles. They come in all shapes and sizes and live in many different habitats, or natural homes. However, all reptiles fit into four main groups: squamates, tuataras, turtles, and crocodilians.

LEARN MORE

LIZARDS AND SNAKES ARE PART OF THE SQUAMATE GROUP. TORTOISES ARE PART OF THE TURTLE GROUP. ALLIGATORS, CAIMANS, AND GAVIALS ARE PART OF THE CROCODILIAN GROUP.

turtle

tuatara

snake

crocodile

A Reptile's Life

Although reptiles differ across the many species, they all share some things in common. A life cycle is the series of basic steps an animal goes through as it grows and changes during its life. Most reptiles go through a similar life cycle.

LEARN MORE

AN ANIMAL'S LIFE CYCLE BEGINS AT BIRTH. IT ENDS WHEN THE ANIMAL IS FULLY GROWN AND READY TO REPRODUCE, OR MAKE BABIES OF ITS OWN.

Snake Life Cycle

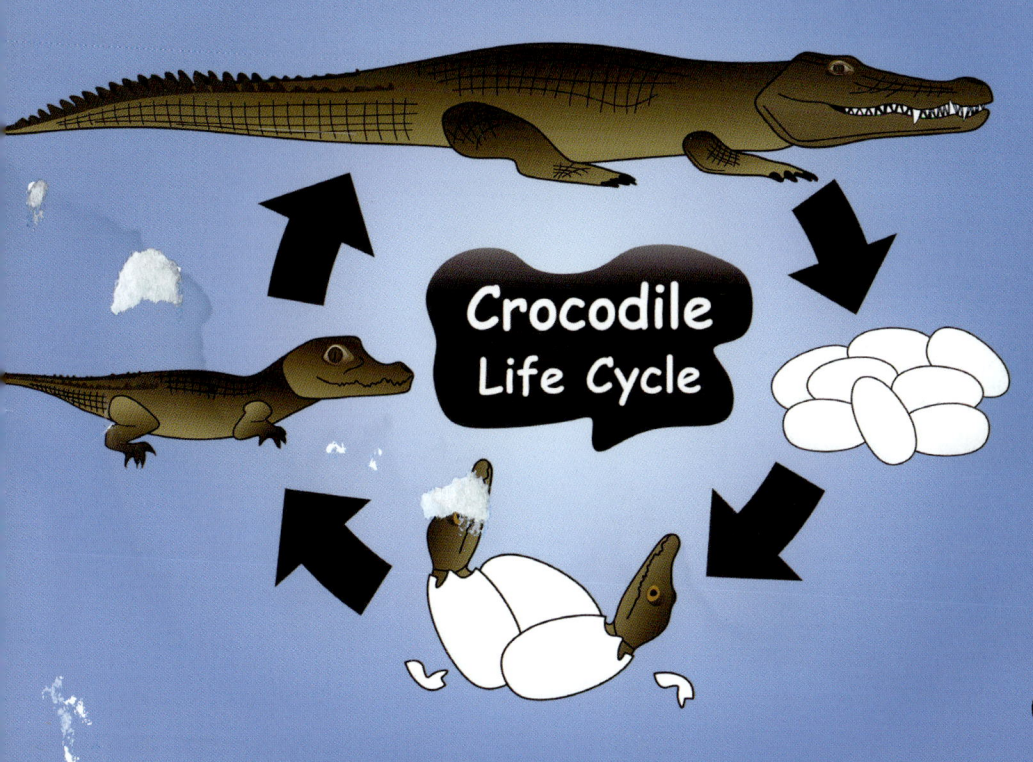

Crocodile Life Cycle

Laying Eggs

Most reptiles lay eggs. The eggshell may be hard or soft and leathery. Females often lay their eggs near water or in soil that's a bit wet. The number of eggs that are laid depends on the species.

LEARN MORE

A GROUP OF REPTILE EGGS IS CALLED A CLUTCH. SEA TURTLES CAN LAY OVER 100 EGGS! LIZARDS USUALLY LAY ONLY ONE EGG.

11

Incubation

After eggs are laid, they need to incubate, or keep warm in order to **develop**. An egg's temperature controls whether the baby will be male or female! In turtles, lower temperatures make female babies and higher temperatures make males. This is also true of American alligators.

> **LEARN MORE**
>
> THE LENGTH OF INCUBATION DEPENDS ON THE REPTILE SPECIES. TUATARAS HAVE THE LONGEST-KNOWN REPTILE INCUBATION. IT LASTS 13 TO 15 MONTHS.

13

Hatchlings

After incubation, reptiles hatch, or break out of their eggs. The babies are called hatchlings. Many reptile hatchlings are independent, meaning they take care of themselves without help from a parent. However, some reptile mothers, such as crocodiles, care for their young.

LEARN MORE

SOME REPTILES USE A SPECIAL BODY PART CALLED AN EGG TOOTH TO BREAK OUT OF THEIR EGG. THEY MAY SHED THIS TOOTH AS THEY GROW OLDER.

15

Juveniles

Young reptiles are called juveniles. Juveniles look like small adults. As reptiles grow, they shed their skin! This **process** is called ecdysis. Snakes shed their skin in one long piece. Lizards, turtles, and crocodiles may shed skin in patches.

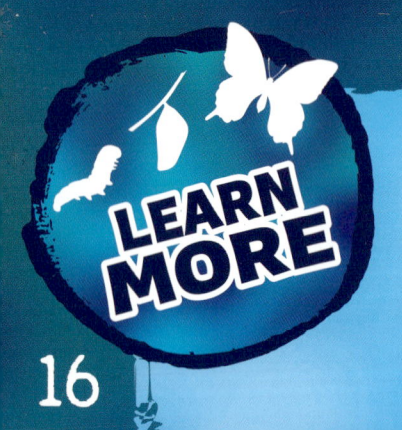

LEARN MORE

REPTILES MAY SHED MANY TIMES AS THEY GROW FROM JUVENILES TO ADULTS. BUT ADULT REPTILES SHED TOO! SHEDDING HELPS REPTILES HEAL THEIR SKIN, CONTROL THEIR BODY TEMPERATURE, AND MORE.

17

Adulthood

Reptiles reach adulthood when they are fully grown and able to reproduce. For some, this takes many years. A reptile's juvenile stage is usually longer if their life is longer. Giant tortoises often live for upwards of 100 years! They reach adulthood around age 25.

LEARN MORE

THE ALDABRA GIANT TORTOISE HAS THE LONGEST LIFE OF ANY REPTILE. ONE NAMED JONATHAN IS THE OLDEST LIVING LAND ANIMAL TODAY! JONATHAN IS OVER 190 YEARS OLD.

Jonathan

Live Birth

Some snakes and lizards are viviparous. That means instead of laying eggs, they give birth to live young! Boa constrictors and anacondas are viviparous snakes. Ovoviviparous reptiles hatch from eggs inside the mother's body! Slowworms are ovoviviparous reptiles.

LEARN MORE

SLOWWORMS LOOK A LOT LIKE SNAKES, BUT THEY AREN'T! THEY'RE REALLY LEGLESS LIZARDS. UNLIKE SNAKES, SLOWWORMS HAVE EAR OPENINGS AND EYELIDS.

snake

slowworm

Life of a Green Anaconda

During spring, adult anacondas come together to **mate**. They may mate in a group called a "breeding ball." The baby anacondas grow inside their mother's body. She gives birth to a litter of as many as 82 young, which are called snakelets.

LEARN MORE

THE TIME WHILE A BABY DEVELOPS INSIDE ITS MOTHER'S BODY IS CALLED A GESTATION PERIOD. A GREEN ANACONDA'S GESTATION PERIOD LASTS 6 TO 10 MONTHS.

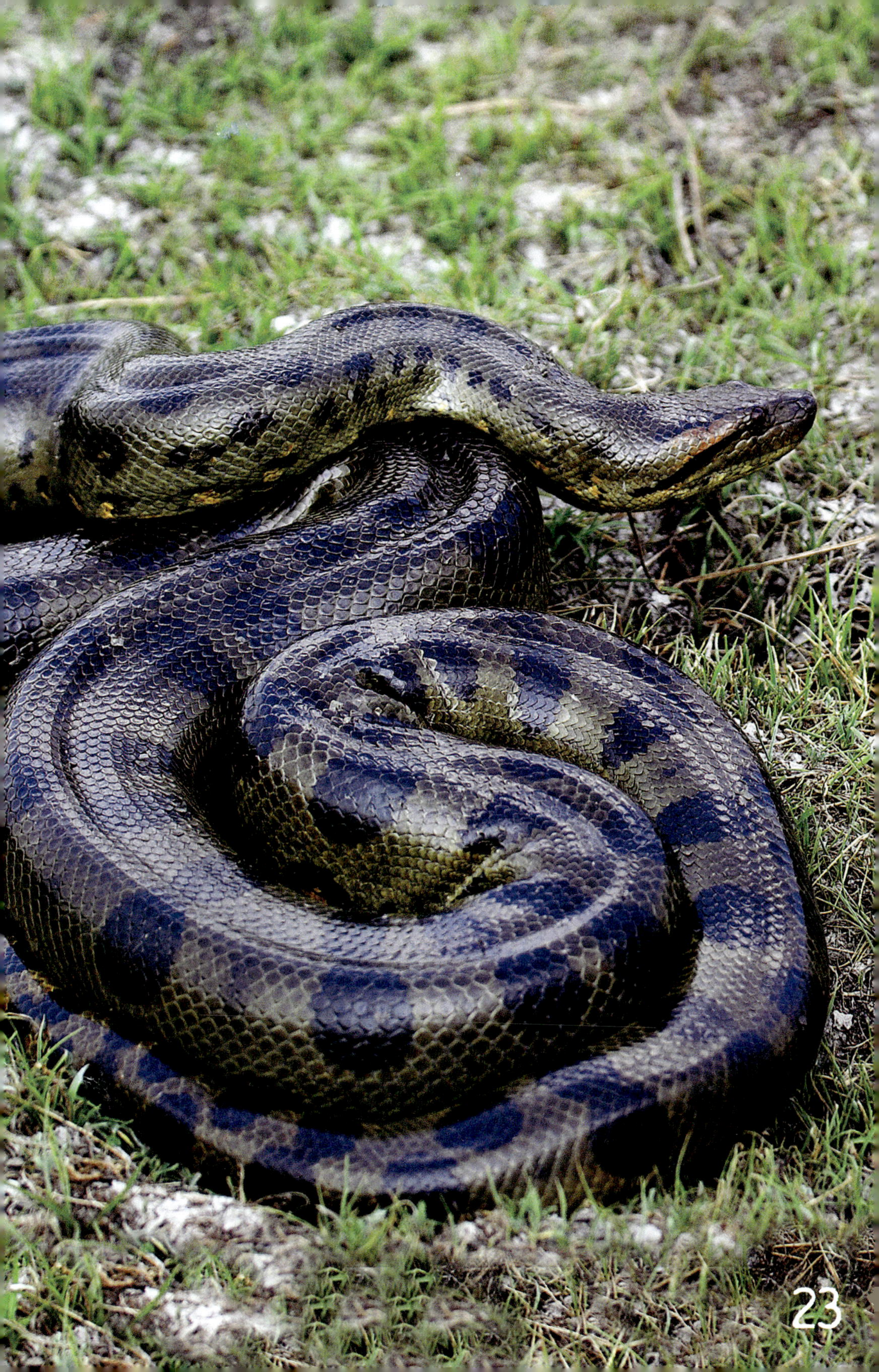

Aquatic and Semi-Aquatic

Although reptiles breathe air, many species spend a lot of time in the water. Reptiles such as sea snakes are fully aquatic. They spend their whole lives in water. Other kinds of reptiles are semi-aquatic. They spend part of their lives on land and part in water.

LEARN MORE

CROCODILES, ALLIGATORS, AND SEA TURTLES ARE SEMI-AQUATIC REPTILES. THEY HAVE SPECIAL **ADAPTATIONS** THAT HELP THEM MOVE AND LIVE IN WATER.

25

A Sea Turtle's Life

Mother sea turtles leave the ocean to lay their eggs on the beach, in a hole in the sand. About 6 to 8 weeks later, babies hatch, climb out, and head to the ocean. Young sea turtles live in the open ocean for about 10 years.

LEARN MORE

SEA TURTLES **MIGRATE** LONG DISTANCES TO FIND THEIR MATES IN SPECIAL AREAS. ONCE A MOTHER SEA TURTLE HAS LAID ALL HER EGGS, SHE GOES BACK INTO THE OCEAN.

Sea Turtle Life Cycle

eggs

hatchling

juveniles

adult

27

Staying Warm

Some reptiles migrate to stay warm. Since they're cold-blooded, reptiles depend on their surroundings to keep their body at the right temperature. They must move to warmer or cooler places based on what they need. Some reptiles are inactive during very cold or very hot times.

LEARN MORE

MANY REPTILES LIKE TO BASK, OR WARM THEMSELVES, IN THE SUN. THIS IS CALLED THERMOREGULATION. TAKE IN HEAT FROM

The Basic Reptile Life Cycle

A female reptile lays eggs. Some lay single eggs, others lay a large clutch.

The eggs incubate under sand or soil.

Babies hatch out of eggs and are called hatchlings. Some are independent at birth. Others are cared for by their mother.

Juveniles eat and grow. They shed their skin as their body gets bigger.

Reptiles reach adulthood when they're fully grown and able to reproduce.

Glossary

adaptation: A change in a type of animal that makes it better able to live in its surroundings.

develop: To grow and change.

mate: To come together to make babies.

migrate: To move to warmer or colder places for a season.

process: To move something forward in a set of steps. Also, the set of steps itself.

temperature: How hot or cold something is.

For More Information

BOOKS

Hughes, Sloane. *20 Things You Didn't Know About Reptile Adaptations*. Buffalo, NY: PowerKids Press, 2023.

Humphrey, Natalie. *Saltwater Crocodile: Giant Reptile*. Buffalo, NY: PowerKids Press, 2024.

WEBSITE

National Geographic Kids
kids.nationalgeographic.com/animals/reptiles/
Learn more fun facts about reptiles!

Publisher's note to educators and parents: Our editors have carefully reviewed this website to ensure it is suitable for students. Many websites change frequently, however, and we cannot guarantee that a site's future contents will continue to meet our high standards of quality and educational value. Be advised that students should be closely supervised whenever they access the internet.

Index

alligators, 6, 12, 24

crocodiles, 14, 24

dinosaurs, 4

egg tooth, 14

giant tortoise, 18

lizards, 6, 20

reproduction, 8, 18, 30

scales, 4

shedding, 16, 30

snakes, 6, 20, 22, 24

squamates, 6

tuaturas, 6, 12

turtles, 6, 9, 10, 12, 16, 24, 26